JOHN GILLOW

printed and dyed textiles from

africa

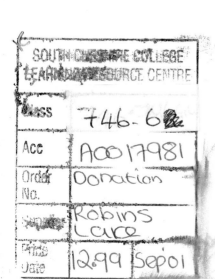

THE BRITISH MUSEUM PRESS

John Gillow has asserted the right to be identified as the author of this work.

First published in 2001 by The British Museum Press
A division of The British Museum Company Ltd
46 Bloomsbury Street, London WC1B 3QQ

A catalogue record for this book is available from the British Library

ISBN 0-7141-2740-X

Commissioning Editor: Suzannah Gough
Designer: Paul Welti
Cartographer: Olive Pearson
Origination in Singapore by Imago
Printing and binding in Singapore by Imago

COVER: Detail of indigo-died *adire* cloth; probably by the Yoruba people, Nigeria. (See pages 80-1)
INSIDE COVER: Imported cotton cloth with indigo tie-dye, probably from Senegal.
PREVIOUS PAGES: Detail of a tie-dyed raphia overskirt from the Congo. (See pages 22–5)
THESE PAGES: Details of a mud cloth by the Bamana people, Mali. (See pages 40-1)

contents

introduction

Africa is a great and varied continent of wide horizons and clear blue skies, which has long held a fascination for those born outside its bounds. Over the centuries its wealth of minerals, animal products and manpower has drawn in colonists and traders, slavers and missionaries alike. Its huge population is of diverse origin: people of Arab and Berber descent in the north, Khoisan-speakers and European colonists in the extreme south, Nilotic-speaking peoples in the north-east, and south of the Sahara a rich mix of groups who speak one of the Bantu languages.

Although the African textile tradition attracted little foreign academic interest until the twentieth century, African textiles found their way into European collections long before then. At the Ulm Museum in Germany, for example, there have been garments made of strip-woven cloth, the characteristic weave of West Africa, since the 1650s, and woven raphia cloth of the Kongo people (decorated with patterns similar to those of the modern-day cut-pile embroidery of the Congo's Kuba tribe) has been held at the Pitt Rivers Museum in Oxford since the seventeenth century.

To go to any market in West Africa is to experience an assault on the visual senses. The market women will be swathed in brightly patterned wraps—usually wax-printed factory-made batiks of Dutch or local origin. At funerals, however, both men and women wear mantles freshly dyed in sombre colours, which in Ghana are decorated with *adinkra* patterns, block-printed on with calabash stamps at the village of Ntonso (see pp. 44–5).

Although some cloth is still woven at home for personal use, there is a vast textile trade. The European wax prints, based on Javanese batik, are an important trade item throughout sub-Saharan Africa. Damasks from China and Europe are another major import, as it is on these and locally made damasks that much of the tied-resist indigo dyeing is done. And until very recently English 'Manchester prints' were hugely popular and heavily traded.

Traditional West African cloths are often made up of very long, narrow strips of fabric,

cut into appropriate lengths and sewn together selvedge to selvedge. Rolls of this cloth once served as currency in place of money. Veils made from the narrowest strips are heavily indigo-dyed, then beaten to produce a metallic sheen. So highly prized are they by Tuareg men that, weight for weight, they are among the most expensive textiles in the world. Indigo dyeing is widespread in West Africa. The dye pits and vats of Kano and Yorubaland in Nigeria and of St Louis in Senegal are justly famous. Indigo was widely grown and traded until recent times, but now it is increasingly replaced by synthetic dyes, combined, in areas such as Senegal or Mali, with an admixture of gentian violet to achieve the deep, shiny purple-blue that is now so fashionable.

Regardless of the dye used, methods of creating white motifs against an indigo-blue background remain the same. Usually a 'resist' is introduced into the cloth before it is dyed by stitching or tying sections of it with raphia or cotton thread, which is pulled extremely tight to prevent the dye from penetrating the enclosed cloth. During the twentieth century machine-stitched resists became common in Nigeria and the Senegambia region. Originally the machine stitching was done by male tailors, but today — as the demands of commerce increasingly break down occupational barriers between the sexes — it is often women who carry out this work, particularly in Nigeria.

More akin to the Asian tie-and-dye tradition are the multicoloured shawls from Matmata in southern Tunisia, while further

A metal pot, in the village of Ntonso, Ghana, of the kind in which dye for printing *adinkra* cloth is prepared.

south, around Tataouine, the woven *bakhnug* wool and cotton shawls are masterpieces of the dyer's art.

Wax is widely employed as a resist medium for drawing designs on to fabric, and in Nigeria cassava starch is the resist used to produce the charming indigo-dyed *adire* cloths of Yorubaland. Hand-decorated *adire* cloths are traditionally made by women and stencilled versions by men, but Yoruba textile specialist

Robert Clyne informs me that nowadays it is often women who do the stencilling as well (although the men cut out the zinc stencils). King George V and Queen Mary at their silver jubilee in 1935 is but one of the eccentric designs that have been used on *adire* cloths.

In Africa painted cloths have a talismanic significance often associated with hunting or warfare, and shirts bearing Arabic calligraphy were once believed to protect the wearer. However, the Senufo hunting cloths of Côte d'Ivoire have been somewhat devalued by becoming a tourist item. Similarly, the *bogolanfini* 'mud cloths' of Mali are now exported wholesale to the United States.

Madagascar has textiles which show both Indonesian and mainland African influences. Ikat on raphia is unique to the island, and beautiful funeral cloths in plain warp stripes are also characteristic.

BARKCLOTH

In central and parts of East Africa barkcloths were one of the main items of clothing until the colonial period. Unless subject to Moslem social and religious influence people would go about either unclothed or wearing a simple piece of barkcloth around their waist or loins. The African method of making this cloth is very similar to that used in Polynesia, except that a different bark is used. The inner bark of a tropical fig tree (*Ficus roko* or *Urostigma ktshyana* in central Africa, *Ficus natalensis* in Uganda) is stripped from the lower trunk and moistened by soaking, steaming or merely leaving it in a wet place. It is then beaten over a smooth hardwood log, using beaters with grooved or cross-hatched heads made of wood, bone or even ivory. Steady beating can thin out the bark to some five times its original width. However, as the fibres run longitudinally, the length can be extended only slightly. Once beaten, the bark is left in the sun for a few days to oxidize to its characteristic reddish brown. The Baganda people of Uganda are reputed at one time to have made about fifty different types of barkcloth, so widespread was its use as clothing and bedding (see pp. 34–5).

RAPHIA

Once used extensively as a weaving material, raphia is still the main fibre used in the manufacture of certain textiles in parts of the Congo, Angola and Madagascar. It is a grassy fibre extracted from the leaves of *Raphia ruffia* or *R. taedigiria,* palm trees natives to Madagascar that also grow around the fringes of tropical forests in central and West Africa. Mature leaves can grow as long as 15 m (50 ft), but it is only the young leaflets that are used. Once cut from the palm, the soft tissues on their undersides are stripped away with the edge of a knife or peeled off by hand to leave only the upper epidermis. The translucent fibres are dried in the sun, then split with the fingernails, shells or a special comb to produce a silky strand about a metre (3 ft) long.

The name 'raphia' is probably derived from the Malagasy word *rofia.* Madagascar, the huge island lying off the south-east coast of Africa, is home to people of diverse ethnic origins. The dominant highlands are inhabited by tribes of Malay descent, who reached the island in the first millennium AD after an incredibly long sea voyage from what is now Indonesia. The coastal regions have mixed populations with a strong black African element. This ethnic mix has resulted in a textile tradition that has much in common with the rest of the African continent, but also certain distinctive features with parallels in Southeast Asia.

As a rule, Malagasy raphia textiles are of a single colour — usually natural undyed beige — but John Mack (1989) states that in three villages in the highlands the Sakalava have been known to produce warp- ikat-decorated cloths, using a technique closely associated with the Indonesian islands. In this method the raphia warps are tied with resists and then dyed. Upon untying, a pattern is revealed by the undyed area. When woven as a warp-faced textile, a two-colour patterned cloth is achieved, such as the simple one illustrated on pp. 26–7. At one time more complex raphia ikats were woven, utilizing more than one dye bath and further tying and untying. This resulted in a colour scheme comprising the base colour, that of the dyes in which the threads had been soaked and a combination of colours where different dyes had been allowed to mix. Human figures and geometric devices were popular motifs.

One of the few other countries of Africa in which warp ikat is practised is Nigeria. There it is used sparingly by the Yoruba, who incorporate it into strip weaves or women's vertical-loom cloths in longitudinal stripes to add contrasting detail (see pp. 70–71).

The Kuba, a tribal confederation based along the Kasai river in the Congo, are viewed by many as the finest artists in Africa. Working in wood, metal and raphia (see pp. 28–31), they were highly resistant to the colonial trade in cotton cloth. Their raphia textile repertoire includes the manufacture of long dance skirts for both men and women, and squares or rectangles decorated by the cut-pile embroidery technique. The men weave the cloth on a single-heddle overhead loom, but because of the restricted length of the fibres, the maximum size of the fabric is about one metre (3 ft) square. The dance skirts are made up by sewing together six or more panels, and there are many ways of decorating them: with embroidery, cut-pile, pierced work, patchwork or appliqué, or by the addition of cowrie shells or bobbled borders. The various styles of decoration are indicative of social status, sex or membership of one of the component groups of the tribal confederation. One of the most visually striking means of decorating the skirts is by tying or stitching in a resist and then dyeing using a colour palette of black, deep browns and reds obtained from dyes made with locally available materials such as camwood, brimstone, or mud and charcoal.

INDIGO

The ancient city of Kano in northern Nigeria was for centuries one of the main termini of the caravan trails from the Maghreb across the Sahara and down to West Africa. A bustling marketplace, it attracted people from all over Africa. As in neighbouring Sokoto, goat hides were tanned there to produce 'Moroccan' leather, and there were also blacksmithing and jewellery quarters. But the most important industry was indigo dyeing. Indeed, in pre-industrial times Kano is reputed to have provided most of the indigo-dyed cloth worn throughout the Sahara. As in the rest of Africa, this cloth has now largely been replaced by mill-made materials and factory-sewn clothing and the natural indigo by synthetic dyes of many hues. Nevertheless, Kano's Kofor Mata dye pits are still in operation today after more than 500 years. According to local tradition, the dyeing techniques used were originally introduced by the Arabs. There are some 150 pits but no more than fifty in active use, and despite a potential workforce of sixty skilled dyers, on a typical day one might see only about ten at work at any one time. Most are middle-aged or elderly men, but there is still the occasional youth learning the trade.

The pits are 4.5 m (15 ft) or more deep and are filled with a dye bath prepared from a mixture of natural indigo, wood ash and potassium soda steeped in water, which is allowed to soak for twelve days. During this period the mixture is pounded vigorously with long wooden staves morning and night, for about fifteen minutes at a time. After this any extraneous plant matter is removed and the mixture is ready for use. The dye bath is said to last for about a year and can be refreshed by adding further organic matter, such as a yellow pulp from the seeds of a local tree. When the pit is exhausted, the residual sediment is emptied out and tipped on to a slag heap. If necessary, chemicals in the waste can be reclaimed by burning and recycled into the next dye vat.

Resist-dyed cloth production in Kano is confined to rather crude tie-and-dye and some stitched resist. The tying, generally done by women, consists of simply tying knots into the cloth in concentric circles. All kinds of cloth are dyed: hand-woven plain cloth, cloth with a resist, or even a pair of old jeans. After soaking for a while in the dye bath the cloth is lifted out for 30–45 seconds, and this process is repeated until it is the right shade of blue. According to the Kano dyers, a light blue takes an hour and a half of dipping and aerating, navy blue three hours, light black four hours, and a deep blue-black six hours. However, none of the products from Kano that I saw could rival the intense blue-black regularly achieved by the master-dyers of Yorubaland.

Indigo is a substantive dye and so needs no mordant, but it has to be fermented and deoxidized in an alkaline solution to convert the indican (the active dyeing agent) into a soluble form that can be absorbed into the fibres. When the indigo-bearing plant matter is soaked in water, enzymic hydrolysis transforms the indican into indoxyl ('indigo white') and glucose. On exposure to the air, the indigo white reoxidizes to give indigo blue —

which is why the cloth is repeatedly exposed to the air during dyeing. Indigo has always been the most popular natural dye: the plants are readily available, it has the great advantage that it dyes cold, and of course indigo-dyed clothes looked wonderful against a dark skin. However, it is not the only dye produced from natural sources. Cassava root yields a natural red dye called *alari*, also said to be obtainable from husks of guinea corn (millet or sorghum). The term 'alari' derives from Arabic *al-hareen*, which refers to red waste silk historically imported from North Africa.

Nigeria is a fertile and wealthy country and the most populous in Africa. Three main ethnic groups predominate: the Yoruba in the south-west, the Igbo in the south-east and the Hausa in the north. All have a strong indigo-dyeing tradition (as do many smaller Nigerian tribal groupings), but it is among the Yoruba that this tradition is the most prevalent. Abeokuta, Oshogbo, Ede and formerly Ibadan in Yorubaland are all famed for their indigo dyeing, but there are many other centres, too.

In Yorubaland, where dyeing has always been a specialized occupation, there is a strong tendency to keep dyeing recipes and methods secret from inquisitive outsiders. There are historical reasons why skilled dyers have settled in particular areas, but in general they are to be found where there is an abundance of indigo-bearing plants or in locations that are on major trade routes — as are St Louis, Kaedi and Kayes, all situated on the Senegal river, which wends its way down from Mali through

Dyeing with indigo at the Kofor Mata dye pits in Kano, Nigeria.

Mauretania and Senegal. Some centres have a tradition of elaborate pattern dyeing, while others restrict themselves to simple designs or none at all. Some of the cloth dyed in Nigeria is hand-spun and/or hand-woven, but the vast majority is now mill-made.

TIE-DYEING

The Dida live on the coast of Côte d'Ivoire, where they make their living by fishing. For ceremonial occasions they plait strands of raphia into skirts, cloaks and kerchiefs (see pp. 32–3), patterning them by means of tied and stitch resist and dyeing them with cola nut and other natural dyes that give a brighter and more varied palette than that of the Kuba. Red and black on a yellowish ground is preferred. Adams and Holdcraft (1992) state that the yellow is obtained from the roots of a shrub and the red from the hardened root of a tree, while the black is said to come from a

combination of manganese and leaves. As with all tie-dyed work, the garments are dyed from the lightest colour to the darkest — in this case first yellow, then red, then black. A cloak or tubular skirt will typically be decorated with circles, ovals and rectangles, often combined with distinct areas of dots, formed solely by means of the tie-and-dye technique. Where the black shades into the red this results in a reddish brown colour, but it remains pure black on the fringes of the garment.

The most colourful of all African tie-dyed textiles are those worn by the Berber women of southern Tunisia to protect their clothes from the oil that they put on their hair. Shoulder cloths known as *ketfiya* and *tajira* are made in the village of Matmata and are dyed by tying small bundles of wheat grains into the rectangular woollen cloth. They are dipped into several different dye baths to achieve a multicoloured pattern, green with red and yellow details being a popular combination. The cloth is scrunched up at either end during dyeing, so that its colour gets progressively lighter towards the ends (Spring and Hudson 1995, pp. 114–15). A distinctive feature of *ketfiya* and *tajira* are the lines of cotton stitching that embellish their tasselled borders. Because wool accepts dye far more readily than cotton, when the *ketfiya* and *tajira* are dyed the woollen base cloth takes up the dye but the cotton stitching does not, remaining its original white. In Africa this exploitation of the difference in dye take-up between wool and cotton is unique to Tunisia (see pp. 50–55), where it finds its highest expression in the supplementary weft-decorated mantles (*bakhnug*) of the Berber women. The ground cloth is wool and the fine supplementary-weft details cotton. Upon dyeing, the cotton details take up the dye less readily than the wool, leaving the fabric almost undyed, in sharp contrast to the deep background colour.

Elsewhere in North Africa it is also among the Berbers that tie-dyed textiles are mostly to be found. In Morocco simple woollen tie-dyed veils are worn by Ayt Atta and Djeballa Berber women (Spring and Hudson 1995, pp. 114–15). In general the craft is characterized by relatively large isolated motifs on thickish woollen fabric, which are created by knotting small stones or pieces of wood into the cloth before dyeing. One of the peculiarities of this craft in North Africa is that the knots may be tied into either side of the fabric. Another is the occasional use of half- or quarter-circle motifs to decorate the edges of the cloth (see pp. 42–3). The fabric is usually dyed only once, the most common colour combinations being black or light blue against a red background, or red or brown against yellow.

Simple tie-dyed cloths are used as food covers in Morocco, Algeria and Libya, as women's belts in Morocco and Tunisia, and as women's head coverings in Morocco and Tunisia. More complex work decorated in a spiral of little dots can be found at Garian in Libya, but this is worked on finer, imported fabric. Mauretania has a tradition of fine tie-

An indigo-dyed cotton cloth decorated with the 'three baskets' design from the Kofor Mata dye pits in Kano, Nigeria.

dyed multicoloured shawls of thin muslin. Throughout Africa tying may be done by pinching up the fabric and tying it off with raphia or cotton thread, but it is common to insert into the ties small objects such as seeds, pebbles, chips of wood, or even buttons.

STITCH RESIST

In the market at Ouagadougou in Burkina Faso one finds stall upon stall selling blue waist and chest wraps patterned with rows of white arrow designs. The base cloth, like much hand-woven fabric in West Africa, is made up of strips 8–10 cm (3–4 in) wide, cut to length then sewn selvedge to selvedge to form a piece eight to ten strips wide. The strips are made by itinerant weavers who carry their portable looms, made of rough-hewn branches, from family courtyard to family courtyard. They will stay weaving in each for as long as they can drag out the task in hand, for they receive food and a present every day, and a parting gift such as a goat when the work is complete.

Women of the Mossi tribe in the north of Burkina Faso take raphia thread and sew tight arrowhead patterns into the cloth. To create a symmetrical design, the cloth is first pleated and then raphia thread is sewn in and out of it and pulled very tight, compressing the cloth so that it will resist the dye.

The Mossi grow both *Indigofera tinctoria* and *Lonchocarpus cyanescus*, the main indigo-bearing plants. The leaves are plucked fresh and steeped in a wood-ash lye in deep, lidded pits set in the ground. Alternatively, balls of dried leaves, either kept back from harvest or acquired through trade, may be used. When the dye pits or vats are ready according to taste, smell and/or look, the resist-sewn cloth is dipped in and stirred around. When lifted out it is a pale white with the faintest green tinge, but slowly this 'indigo white' reoxidizes to blue. When the desired deep blue has been achieved (which may take many dippings), the raphia threads are snipped off, or the cloth just pulled laterally in a steady movement to 'pop' off the ties, and the resist pattern is revealed. Raphia is preferred to cotton thread as it is tougher and thicker, and easier to remove. The Yoruba sometimes stretch an intricately tied piece over their thighs and scrape the ties off with a razor blade, but this risks making holes in the base cloth.

Similar cloths are created by this method and variations of it all over West Africa, from Senegambia in the north to Cameroon in the south, although in many places natural indigo has been replaced by synthetic. West African stitch resist is almost always blue or brown. Cola nut produces brown, but indigo blue is by far the most popular colour. In Nigeria, Gambia, Senegal, Mali and other parts of West Africa, stitch-resist textiles are produced by machine sewing as well as by hand.

Imports of fine European mill cloth and smooth sewing threads have also meant that more delicate work can now be done, and probably the finest examples of stitch resist ever produced are the indigo-dyed cloths made in St Louis, in northern Senegal on the Mauretanian border, before World War II. European mill-woven cloth was folded and

Tying resists into leather at Oshogbo in Yorubaland, Nigeria.

pleated and then sewn with cotton thread in patterns so intricate as to merit being called embroidery. After dyeing in indigo, the thread resists were painstakingly unpicked to reveal Moorish-inspired patterns, probably derived from nearby Mauretania. This tradition has since been revived, employing the same method but using rather harsh blue or brown synthetic dyes and with the embroidery resists adapted to modern Senegalese taste (see pp. 58–9). The main pitfall is again that because the thread used is cotton rather than raphia, unpicking it is hazardous and great care must be taken not to damage the cloth.

Etu is the Yoruba name for the guinea-fowl pattern, which can be applied to any textile design reminiscent of the speckled plumage of this bird. The Yoruba produce beautiful tied-resist and stitch-resist *etu* cloths in square and circular patterns, which are so pleasing to the eye that often they are left untied or unstitched.

In Yorubaland the resist-patterned *adire* cloth is made in two principal ways: using a tied or stitched resist, or by the starch-resist method. Tie-dyed cloth is known as *adire oniko*, a term derived from the Yoruba word for raphia, and stitched resist as *adire alabere*, from the word meaning stitch. The term for starch-resist fabric is *adire eleko*.

The *adire alabere* stitch-resist method is the same as the Indonesian *tritik* technique and involves either stitching by hand with raphia thread or machine sewing with cotton. The cloth may first be folded or pleated before

being stitched ready for dyeing, which results in a variation or repeat of the basic pattern. To give further variety, the cloth is sometimes pleated longitudinally or diagonally. The stitching (either running stitch or hem stitch on the edges of the pleats) may lie flat or may be pulled tight. In the latter case the thread needs to be strong — another good reason for using raphia. Successive folds of the fabric are gathered up tightly against each other in such a way that the fabric is sufficiently compressed to form a resist against penetration of the dye. Hand-stitching of *adire alabere* is traditionally done by women and machine sewing by men, although such gender differentiations are becoming increasingly blurred over time.

The Igbo of Nigeria produce stitch-resist pictorial cloths known as *ukura*, which often feature leopard and other animal motifs. These are traditionally made in northern Igboland for the Leopard Society of Cross River in the extreme south-east of Nigeria (see pp. 74–5).

STARCH RESIST

Adire eleko cloth is a speciality of Yoruba dyers. Their starch-resist technique works on the same principle as wax-resist batik and is well suited to indigo dyeing, in that the temperature of the vat never rises high enough to dissolve the starch (which is not an overly strong resist). The starch used — usually cassava or cornflour — is known as *lafun*. It is mixed with alum and can then be applied in either of two ways. The first is the original method, used by women, of applying the starch with a palm rib or bird's-feather quill (see pp. 72–3). The traditional centre for this type of production was Ibadan. In the second, more recent method — probably dating from the last decades of the nineteenth century — the *lafun* is applied through metal stencils (see pp. 76–7). At first this was done only by men, but the situation is now changing. The main centre for this method is Abeokuta.

Hand-drawn *adire eleko* production is highly labour-intensive. The patterns used are generally traditional ones handed down with very little change from mother to daughter. The woman first draws out a grid in *lafun* and then proceeds to fill in the squares with a variety of motifs. She will often put her own personal mark on the hem of the cloth in order to identify it . The amount of *adire* cloth that can be decorated by hand in a day depends on the complexity of the design. Stencilling, on the other hand, is much faster and an impressive amount of yardage can be produced in the same period. In both cases the *lafun* is applied to only one face of the cloth.

Adire eleko stencils were once made out of the linings of tea chests or cigar boxes but are now chiselled out of 30 × 20 cm (12 × 8 in) rectangles of zinc. The fabric to be stencilled, which may be either plain white cloth or coloured and patterned mill cloth, is nailed to a worktable and the *lafun* is pressed through the stencils with a metal spatula. Any surplus is retained for further use. Most cloths require

a series of stencils, which are used in descending order of their importance to the design. In the case of both hand-drawn and stencilled *adire eleko* cloth, repeated immersion in the indigo dye bath is required — a process over which great care is taken. When all the dyeing is complete the *lafun* is scraped off the cloth and it is hung out to dry, after which it is deemed ready to wear. Although the cloths are rinsed, since water is a scarce commodity some may retain excess indigo, which can come off on the skin. However, this is thought by some to have therapeutic powers. It should be noted that although stencilled *adire eleko* production is centred in Abeokuta, it represents only a very small proportion of the town's total textile output. There are many different workers making such products as machine-sewn and clamp-resist dyed textiles in a wide range of colours other than the traditional blue.

WAX RESIST

The use of wax as a resist in combination with synthetic dyes is very common in Africa. Applied by dipping a sponge into molten wax and squeezing it by hand on to the cloth to be dyed, it is both easier to apply and more durable than such traditional resists as starch.

A zinc stencil used for decorating starch-resist *adire* cloth from Oshogbo, Yorubaland, Nigeria.

The wax is boiled out of the cloth after dyeing and normally reused. Most of the variations seen in traditional and modern batik in other parts of the world are also to be found in Africa. The crumpling of the wax resist before dyeing to give a crackled effect in the finished cloth has a history that stretches back at least to the nineteenth century. A brightly coloured 'psychedelic' look is most popular, particularly among the young.

CLAMP RESIST

Although clamp-resist dyeing is associated principally with Japan, it is also used to some extent in West Africa. The cloth to be dyed is first pleated and folded into a cube, then vertical pressure is applied to it, either by tying very tightly in the form of a cross or by applying a clamp. When the compressed cloth is dipped into the indigo dye bath, only its outer folded edges are penetrated by the dye. Upon unfolding, a pattern of squares is revealed in white (or whatever was the colour of the base cloth) against a blue background.

ADINKRA

Ghana is one of the very few countries in sub-Saharan Africa where block printing of cloth is carried out (another notable instance being among the hunters of Sierra Leone). In Ghana, as in many parts of Africa, funerals are of great symbolic importance and mourners are obliged to dress in dark, sombre colours. In the village of Ntonso, close to Bonwire (the centre of Ashanti weaving) and the great market town of Kumasi, one can see elderly men dyeing cloth for this purpose (few young men are now taking up this poorly paid work). For many people it is sufficient to take an old, brightly coloured strip-weave *kente* cloth and dye it black in an infusion of the bark of the badee tree, but those so inclined will commission a special wrap of *adinkra* hand-printed cloth, traditionally associated with mourning.

As the base fabric for *adinkra* cloths the skilled textile printers use lengths of Chinese mill cloth measuring four yards by three (approx. 3.5×2.75 m). On to this they print moon, fern and many other traditional motifs (each of which has its own symbolic meaning) using a stamp carved out of a calabash gourd. The design motif is carved into the hard outer surface of the gourd and a handle is made by pressing four raphia-palm splints into its soft inner skin and drawing their ends together. The craftsman draws out a grid on the cloth with another splint dipped into a thick, dark goo made by boiling down root bark, again from the badee tree. He then applies rows of a different set of design motifs to each square of the grid by repeatedly rolling one of the slightly curved stamps within that area. Sometimes he may decorate alternate squares with parallel lines by drawing a small bamboo comb across them. Each man can complete about two *adinkra* cloths a day, which are then hung out overnight to catch the dew.

Adinkra cloths made for funerals and mourning are overdyed red or black, but others retain their white background and are worn on festive occasions. In today's *adinkra* cloths the rows of printed squares are divided by longitudinal lines using a type of faggoting stitch in red, black, yellow and green. In the past a narrow woven strip in the same colours could be used for the same purpose.

BOGOLANFINI

The *bogolanfini*, or 'mud cloths', of Mali (see pp. 36–41) are some of the most striking and popular of all African textiles. Made by Bamana/Bambara women to the north of Bamako, Mali's capital, their original use was as hunters' shirts or women's wraps, but post-independence they have become a fashion item — not only within Mali itself but right across Africa and beyond. Much of Mali's *bogolanfini* production now heads straight for New York.

Traditional mud cloths are decorated with geometric patterns in white on black, and the decorative process is of some interest. At first glance it would appear to be a resist technique, but this is not the case. Beautifully hand-woven strip-cloth is first soaked in a mordant-bearing mulch of leaves from local trees such as *Anogeissus leiocarpus* (*n'galaman*) and *Combretum glutinosum* (*n'tjankara*). This impregnates it with the mordant (tannin), but at the same time turns it a deep urine yellow. Designs are drawn in outline on the cloth in river mud that has been kept for a year or more and is rich with iron salts. Working carefully around the outlined motifs, the remainder of the cloth is then covered with the mud, using a blunt knife or spatula or even a toothbrush.

Carving an *adinkra* stamp out of a calabash gourd, in the village of Ntonso, Ghana.

The iron oxide in the mud reacts with the tannic acid in the cloth to produce a colourfast black 'background' to the design. Finally the mud is washed off and the yellow design areas bleached back to the cloth's original white with a bleaching agent of millet bran and peanuts, whose active ingredient is caustic soda.

Around Mopti cruder versions of *bogolanfini* are now produced, in which no bleaching back is done and the final result is a few scattered black motifs set against a background of dirty yellow, or rust from a wild vine (Balfour-Paul 1999).

PAINTED CLOTHS

Since the very beginnings of Islam, the writing down of quotations from the Koran or the Hadiths has been regarded as having protective powers. With progressive Islamicization, the Muslim custom of inscribing textiles and clothing with such texts spread down into black Africa from the heartlands of Islam in Arabia, Asia and North Africa. Shirts not only covered in Arabic calligraphy but decorated with leather amulets containing yet more Koranic quotations became greatly prized by tribal warriors. The famous Hausa trading city of Kano developed as a major centre for their production and distribution, and from there Hausa craftsmen moved on to other parts of West Africa, where they made similarly inscribed cloths for Islamic and other groups.

One major non-Islamic group for whom they created 'calligraphy cloths' were the Ashanti in what is now Ghana. The Ashanti had a strong warrior tradition and were no doubt proud to embrace an Islamic custom that celebrated valour. These cloths were laid out in a grid of tiny squares, some of which were filled with Arabic-style writing, while others were painted with designs in black, red, green and yellow (see pp. 46–9). However, on few such cloths does the written Arabic make any real sense — a trait held in common with calligraphy cloths found in other parts of the world. David Heathcote, who researched the matter in Kano in the early 1970s, states that the men who drew out these cloths were the same ones that drew out embroidery patterns.

THE FUTURE

Throughout Africa there is a continuing trend for locally woven traditional fabrics to be superseded by modern factory-made cloth, preferred for its bright colours, washability, and ease of tailoring into Western-style fashionable clothes.

At the same time two important factors are helping to preserve the legacy of African hand-crafted cloth-making, at least in the short term. One is that traditional garments are still deemed essential for ceremonies marking 'rites of passage', and for funerals in particular. The other is that certain traditional textiles are becoming increasingly sought after in the West. This in turn influences the fashion tastes of the urban elite in African countries, whose traditional weavers, dyers and embroiderers will continue to produce fine cloths for as long as there is local demand for them.

Export orders and tourist-oriented production go a long way towards keeping Africa's craftspeople in work, but if the discerning domestic market declines, then inevitably, so too will standards. However, traditional customs and beliefs are in many places still very strong, and the textiles required to maintain them will — it is to be hoped — keep African textile traditions alive for the foreseeable future.

MOROCCO

ATLAS MTS.

TUNISIA

B e r b e r

S A H A R A

T u a r e g

MALI

Senegal

SENEGAL

Niger

B a m b a r a

Bamako ◆

Mossi
Ouagadougou ◆

• Kano

H a u s a

NIGERIA

Senufo

Dida

GHANA

Ashanti

Ewe

Yoruba

Igbo

CAMEROON

Nile

*ETHIOPIAN
HIGHLANDS*

Congo

UGANDA

Baganda

DEMOCRATIC
REPUBLIC
OF CONGO

Lake Victoria

K u b a

*INDIAN
OCEAN*

*ATLANTIC
OCEAN*

S a k a l a v a

MADAGASCAR

◆ national capital

• city / town

Ewe people

LAIMASAKA
Warp ikat-dyed raphia woven mantle.
3.22 m × 98 cm (10 ft 6¾ in × 3 ft 2½ in)

A BOLD OVERALL EFFECT IS TONED DOWN
BY THE SUBTLE USE OF NATURAL COLOURS.
THE INCLUSION IN THE DESIGN OF
MOUNTED HORSEMEN AND BUFFALO (SEE
DETAILS OVERLEAF) REFLECTS THE INDO-
NESIAN HERITAGE OF THE MOUNTAIN
MALAGASY, WHILE THE MOTIF ON THE
RIGHT IS REMINISCENT OF THE UNION
JACK, SEEN AS A SYMBOL OF POWER.

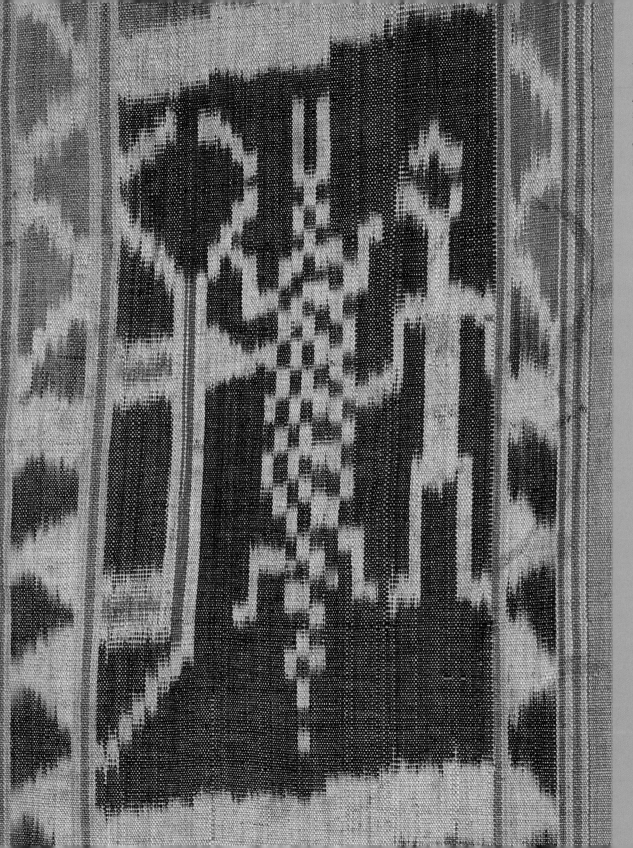

THESE THREE MOTIFS —
HUMAN FIGURE, CROCODILE
AND BUFFALO — SIGNIFY
STRENGTH AND FERTILITY.
BELIEVED TO PROTECT THE
WEARER, THEY ALSO DENOTE
PRESTIGE.

THE MOUNTED HORSEMAN
IS ANOTHER POWERFUL
MOTIF DRAWN FROM THE
TRADITIONAL WARP-IKAT
TEXTILES FROM THE
INDONESIAN ISLANDS.
AMONG THE SAKALAVA
PEOPLE OF MADAGASCAR
IT HAS ARISTOCRATIC CON-
NOTATIONS AND INDICATES
HIGH STATUS.

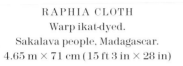

RAPHIA CLOTH
Warp ikat-dyed.
Sakalava people, Madagascar.
4.65 m × 71 cm (15 ft 3 in × 28 in)

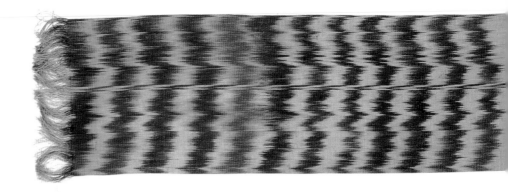

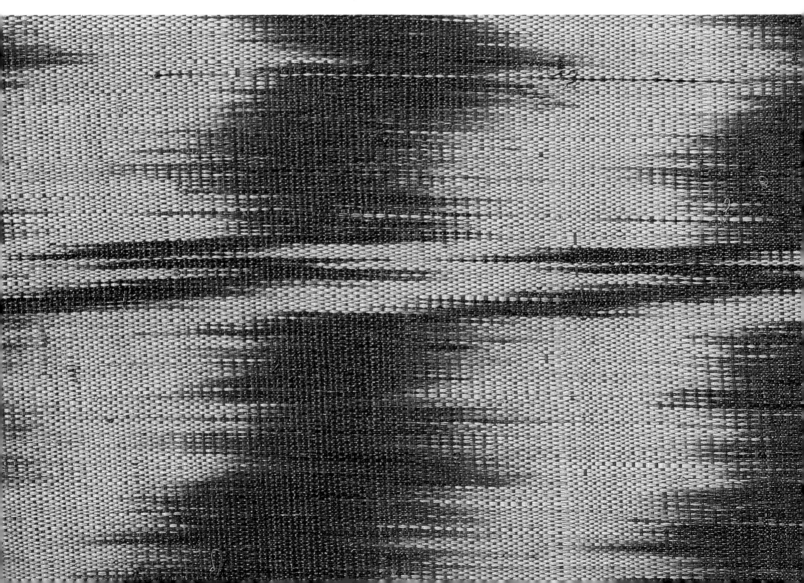

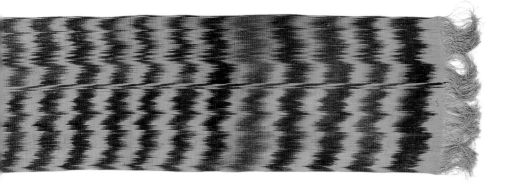

THE IKAT TECHNIQUE PRODUCES AN 'OUT-OF-REGISTER' EFFECT, BUT THE SIMPLE TWO-COLOUR ZIGZAG PATTERN ON THIS CLOTH MAY HAVE BEEN ENHANCED BY PULLING THE WARPS OUT OF POSITION AFTER DYEING THE THREADS AND UNTYING THE RESISTS PRIOR TO WEAVING.

OVERSKIRT
of tie-dyed raphia
197 × 113 cm (77½ × 44½ in)

MANY OF THE RAPHIA TEXTILES
OF THE CONGO ARE SUBLIME
EXAMPLES OF THE WEAVER'S ART.
HERE THE CENTRAL MOTIFS ON
THE BORDER HAVE BEEN ACHIEVED
BY FOLDING THE CLOTH BEFORE
TIE-DYEING IT.

THE DRAMATIC GEOMETRIC DESIGN
IS THE RESULT OF TYING A REPEAT
DIAMOND PATTERN OF SMALL
DOTS AROUND THE CENTRAL
MOTIF. THE TIE-DYE TECHNIQUE
ALLOWS THE COLOURS TO BLEND
INTO ONE ANOTHER.

OVERSKIRT OR CLOAK
of tie-dyed raphia
184 × 127 cm (72½ × 50 in)

THE CHARM OF AFRICAN TIE-DYED TEXTILES
OFTEN LIES IN AN APPARENTLY ARBITRARY
ARRANGEMENT OF DESIGN ELEMENTS. IN
THIS CASE CHAINS OF A ROUNDED-DIAMOND
MOTIF LOOSELY FORM DIAGONAL CROSSES
AND ARE INTERSPERSED WITH LARGER,
MORE DETAILED MOTIFS THAT REPEATEDLY
ECHO THE SAME BASIC SHAPE.

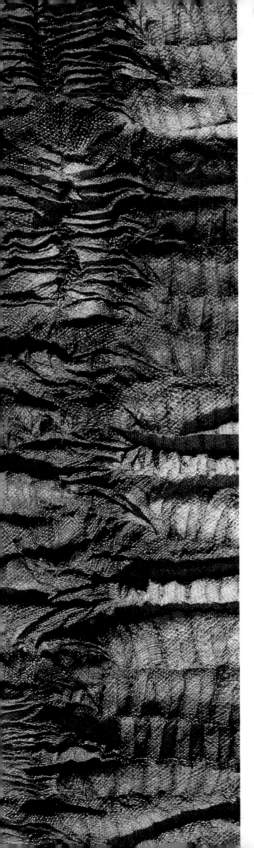

WOMAN'S CEREMONIAL SKIRT
of plaited raphia with
tied and stitched resist decoration
60 × 40 cm (23⅔ × 15¾ in)
Length including fringe: 93 cm (36⅔ in)

THE BOLD COLOURS USED IN THIS DESIGN ARE ENRICHED
BY THE USE OF DEEP FOLDS. THIS CRINKLED TEXTURE IS
LEFT UNSMOOTHED, PERHAPS TO COMPLEMENT THE
CICATRICES (DECORATIVE SCARIFICATION) OF THE WEARER.

BARKCLOTH
with painted decoration
202 × 240 cm (79½ × 94½ in)

BARKCLOTHS WERE MADE IN LARGE NUMBERS IN UGANDA IN THE EARLIER PART OF
THE TWENTIETH CENTURY. THIS EXAMPLE IS PAINTED AND STAMPED TO CREATE A BOLD
PATTERN IN BLACK THAT LEAPS OUT FROM THE DEEP RED-BROWN OF THE BACKGROUND.

BOGOLANFINI ('MUD CLOTH')
probably worn as a woman's wrap
150 × 103 cm (59 × 40½ in)

THE BOGOLAN CLOTHS OF
MALI ARE ESSENTIALLY
PROTECTIVE IN NATURE.
THEIR DISTINCTIVE DENSE
PATTERNING, PAINTED IN
NEGATIVE, IS BELIEVED
TO ABSORB THREATENING
SPIRITUAL FORCES.

THIS FINE EXAMPLE OF
A MALI *BOGOLANFINI*,
OR MUD CLOTH, GAINS
DRAMATIC IMPACT FROM
THE USE OF THE SAME
WIDTH OF LINE THROUGH-
OUT ALMOST THE ENTIRE
DESIGN.

BOGOLANFINI ('MUD CLOTH')
probably worn as a women's wrap
151 × 102 cm (59½ × 40 in)

BOGOLANFINI WERE TRADITIONALLY WORN BY PEOPLE CONSIDERED TO BE IN DANGER, SUCH AS HUNTERS AND YOUNG WOMEN. ANY MALEVOLENT

EXTERNAL FORCES WERE THOUGHT TO BE DRAWN INTO THE WEAVE, WHERE THEY BECAME TRAPPED BY THE COMPLEX MEANDERING PATTERNS.

BOGOLANFINI ('MUD CLOTH')
of the first half of the twentieth century
136 × 104 cm (53½ × 41 in)

MUCH FINER WORK IS EVIDENT IN THIS
MUD CLOTH, DATING FROM BEFORE WORLD
WAR II, THAN IN MODERN EXAMPLES. THIS
DESIGN IS ENLIVENED BY USING A BORDER
PATTERN ALONG TWO SIDES ONLY,
CREATING A POWERFUL ASSYMETRY.

VEIL
of tie-dyed wool
45 cm (17¾ in) square

THE DYNAMICS OF THIS DESIGN RELY ON THE TENSION
INTRODUCED BY THE CIRCULAR MOTIF BEING ALMOST TOO
LARGE FOR THE DIMENSIONS OF THE CLOTH. THE PARTIAL
MOTIFS TIED INTO THE CORNERS ARE TYPICAL OF MOROCCAN
BERBER TEXTILES.

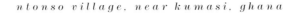

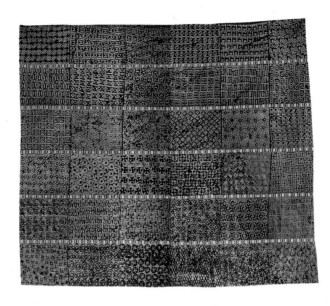

ADINKRA
block-printed man's mourning cloth
1.97 × 3.08 m (6 ft 5 ½ in × 10 ft 1 ¼ in)

RED IS THE COLOUR OF MOURNING AMONG THE ASHANTI. THIS *ADINKRA* MOURNING CLOTH HAS BEEN HAND-PRINTED ON A LIGHT RED GROUND WITH A BARK-DERIVED BLACK DYE. THE MOTIFS ARE PRINTED WITH STAMPS CARVED FROM CALABASHES AND EACH ONE HAS A SPECIFIC RITUAL MEANING.

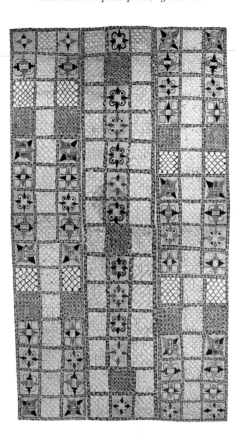

PAINTED CLOTH
probably made by Hausa craftsmen
for the Ashanti in Ghana
3.48 × 2 m (11 ft 5 in × 6 ft 6¾ in)

THIS COLOURFUL TALISMANIC CLOTH HAS A
STRONG ISLAMIC HERITAGE. ITS BOLD BUT
ORDERLY DESIGN INCLUDES MOTIFS
INSPIRED BY ARABIC CALLIGRAPHY (SEE
DETAILS OVERLEAF) — INTERPRETED IN A
STYLE THAT IS UNMISTAKABLY AFRICAN.

46

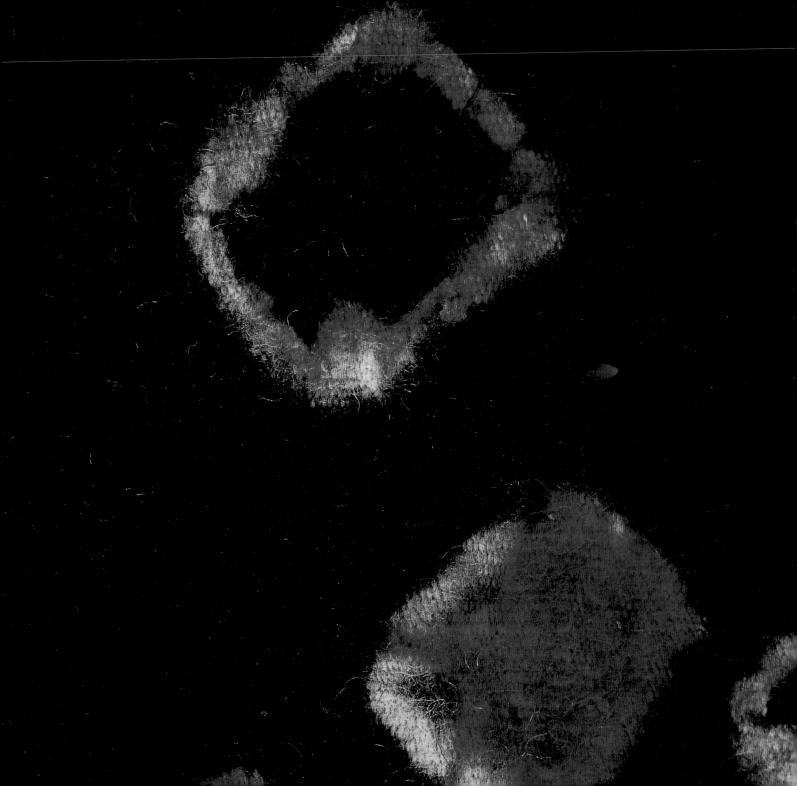

HEADCLOTH
of tie-dyed wool with embroidered top corners
118 × 85 cm (46½ × 33½ in)

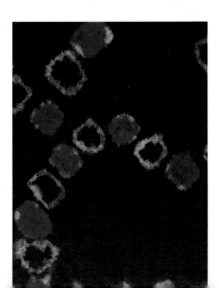

THE TIGHT EMBROIDERED MOTIFS IN THIS DESIGN PROVIDE CONTRAST WITH THE LOOSE EFFECT OF THE TIE-DYEING. THE STITCHERY IS IN COTTON, BUT THE CLOTH ITSELF IS MADE FROM WOOL, WHICH TAKES UP THE DYE MORE READILY. THE USE OF DIFFERENT-COLOURED DYES, THOUGH RARE IN SUB-SAHARAN AFRICA, IS CHARACTERISTIC OF NORTH AFRICA.

BAKHNUG SHAWL
of woven wool with cotton details and fringe
133 × 155 cm (52 × 61 in)

THE PATTERN ON THIS SHAWL IS ACHIEVED USING A
SINGLE DYE COLOUR. THE TECHNIQUE RELIES ON THE
FACT THAT THE WOOLLEN FABRIC WILL READILY TAKE UP
THE DYE, WHILE THE COTTON DECORATION WILL NOT.
AS CAN BE SEEN HERE AND OVERLEAF, THE CLOTH IS
DECORATED WITH A LIMITED ARRAY OF COMPLEX DESIGN
ELEMENTS — A TYPICAL FEATURE OF ISLAMIC CRAFTS.

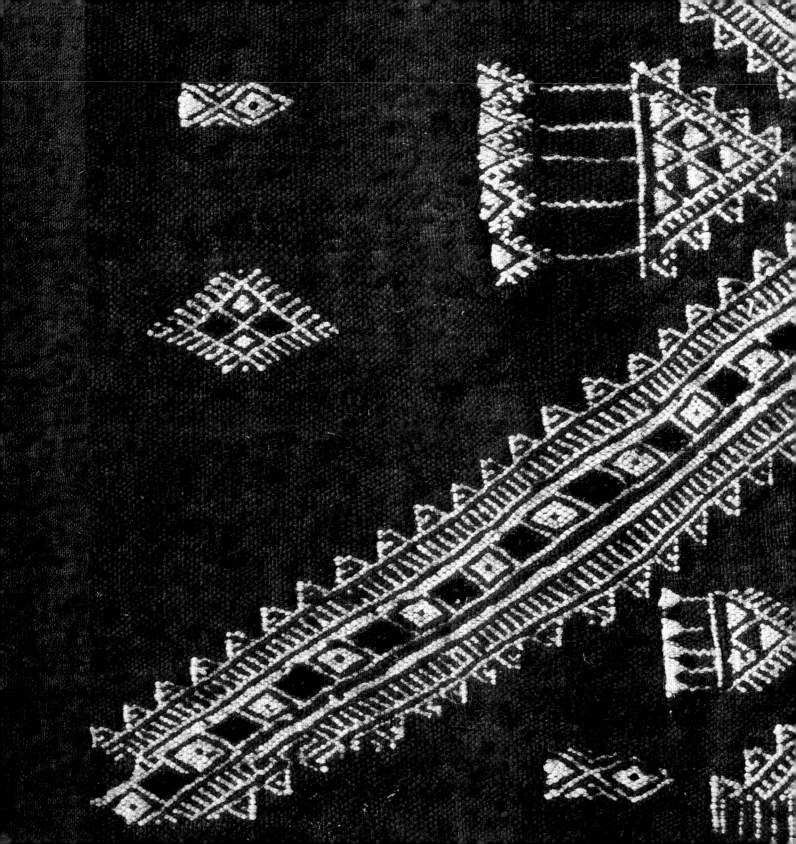

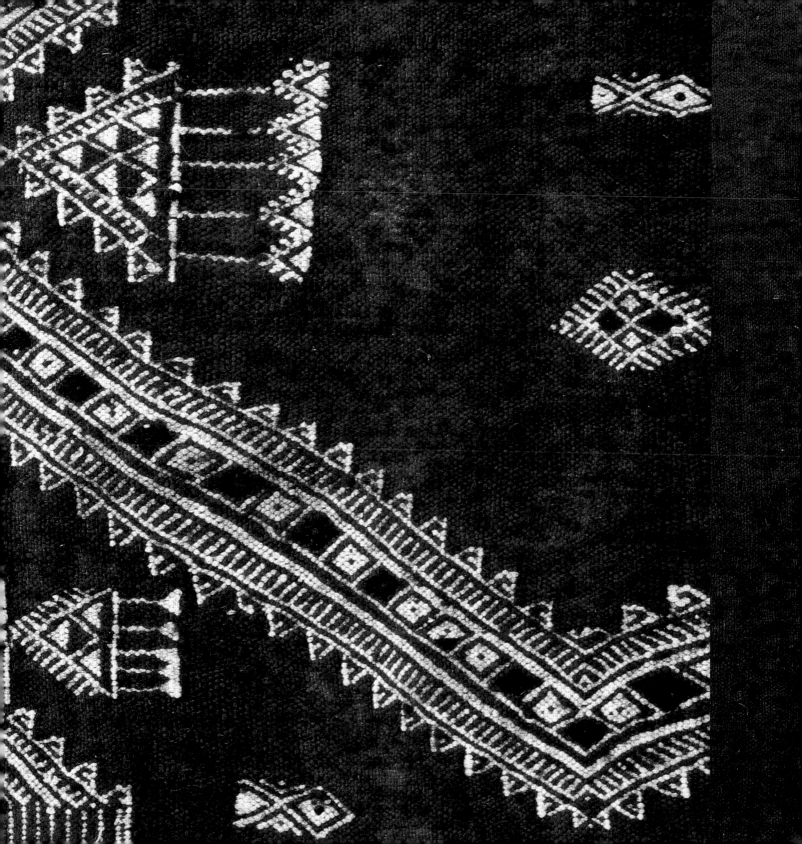

SASH
of tie-dyed wool
198 × 26 cm (78 × 10¼ in)

THE DYNAMIC OF THIS DESIGN,
IS CREATED BY INSERTING
LARGER RINGS AT EITHER END OF
THE CLOTH AND ALSO ALLOWING
THE COLOUR TO CREEP INTO THE
TASSELS.

INDIGO-DYED CLOTH
with resist decoration
233 × 135 cm (91¾ × 53 in)
without fringe

THIS CLOTH HAS A LINEAR ARRANGEMENT
OF EMBROIDERED COTTON RESIST MOTIFS
ALTERNATING WITH SEWN, PLEATED AND
GATHERED RESIST BANDS. THE INTRICACY
OF THE FORMER (WHICH ARE PROBABLY
ISLAMIC-INSPIRED) IS BALANCED BY THE
SIMPLICITY OF THE LATTER.

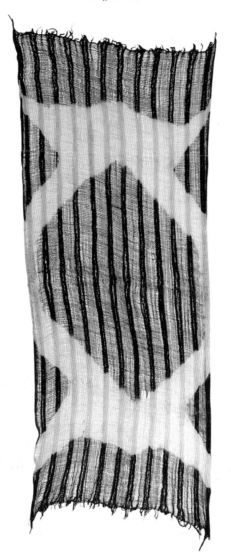

A LARGE, BOLD RESIST-DYED
PATTERN ON LOOSE OPEN-
WEAVE CLOTH, REVEALING

THE NATURAL BEIGE OF A
COTTON VARIETY NATIVE TO
NIGERIA.

SHAWL
mud-resist dyed
204 × 90 cm (80 × 35½ in)

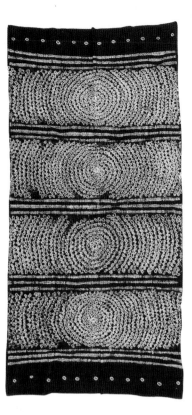

ADIRE CLOTH
indigo-dyed using a combination
of resist techniques
163 × 80 cm (64 × 31 ½ in)

A MIXTURE OF TIE-DYED
AND PLEATED-AND-STITCHED
MOTIFS HAS BEEN USED
HERE TO CREATE A RHYTHM
OF FREE-FLOWING SPIRALS
AND LINEAR BANDS.

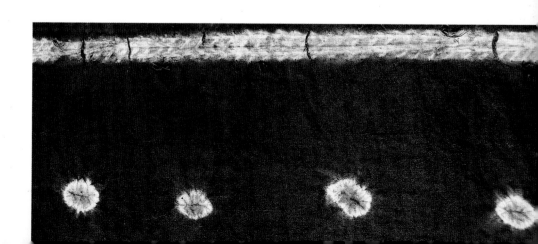

ADIRE CLOTH
tie-dyed with indigo
204 × 167 cm (80 × 65¾ in)

LARGE FLORAL MOTIFS ARE
CONTAINED WITHIN A TIGHTLY
ORGANIZED STRUCTURE OF
VERTICAL BANDS, IN A
MANNER REMINISCENT OF
THE PRINTED DESIGNS ON
IMPORTED MILL CLOTH.

WOMAN'S WRAP
of indigo-dyed *adire* cloth with hand-drawn design
196 × 176 cm (77 × 69¼ in)

THIS ELEGANT DESIGN — SIGNED BY THE ARTIST (SEE DETAIL, LEFT) — ACHIEVES A SUCCESSFUL MARRIAGE OF THE GEO-METRIC WITH THE FIGURATIVE. THE LIZARD (RIGHT), AN ARCHETYPAL TROPICAL SYMBOL, SITS ON A 'POINTILLISTIC' GROUND ENCASED IN SWIRLING TRAILS WHICH MAY PERHAPS SYMBOLIZE SOUND.

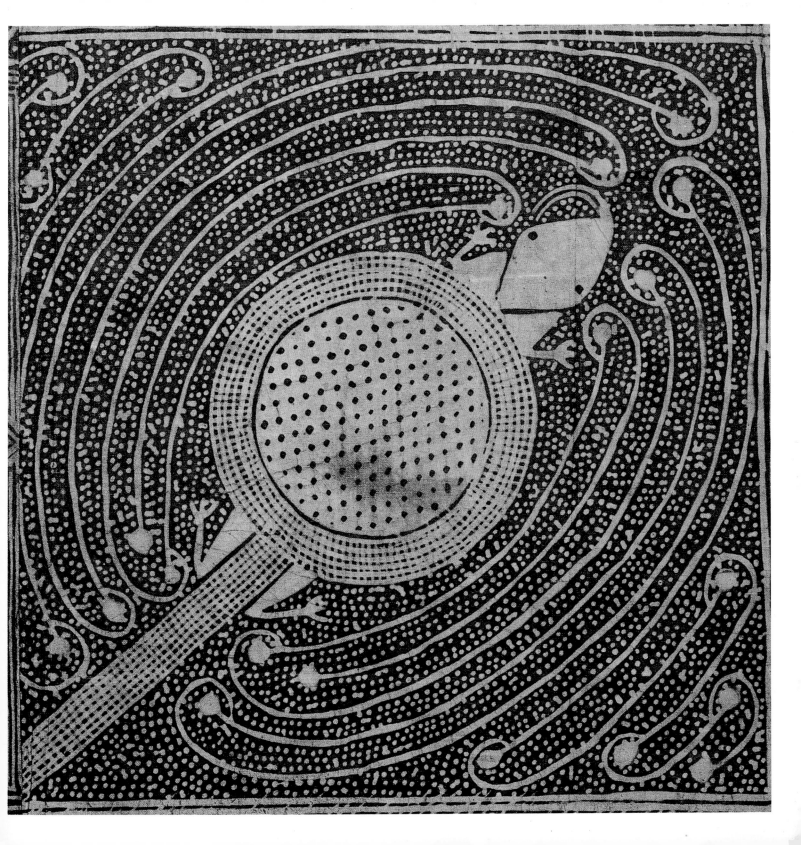

TRACES OF LIGHT BLUE ARE
EVIDENT WHERE THE INDIGO
DYE HAS BEGUN TO PENETRATE
THE STARCH USED TO APPLY
THE DESIGN. THE UMBRELLA
SET WITHIN A DIAMOND IS A
SYMBOL OF ROYALTY.

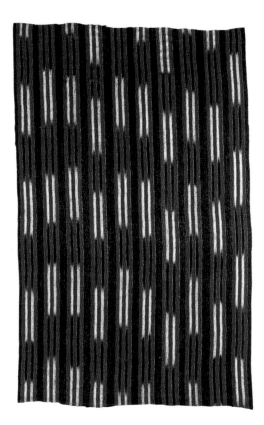

MANTLE
of cotton and silk, incorporating
warp-ikat details
178 × 114 cm (70 × 45 in)

THE YORUBA ARE ONE OF THE FEW WEST
AFRICAN PEOPLES WHO USE WARP-IKAT DYING
TECHNIQUES IN THEIR TEXTILE PRODUCTION.
IN THIS DESIGN THE FORMALITY OF THE WARP
STRIPES, REGULARLY SPACED IN GROUPS OF
THREE, IS OFFSET BY THE FREER ARRANGE-
MENT OF THE ALTERNATE WHITE AND INDIGO
BLUE BANDS THAT FILL THE INTERVENING LINES.

71

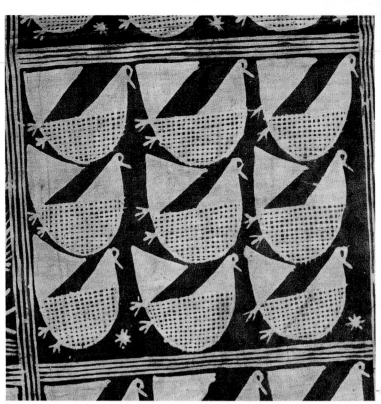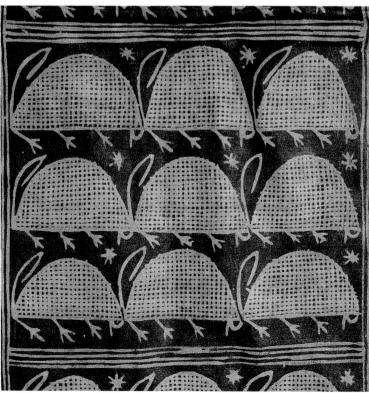

WOMAN'S WRAP
of indigo-dyed *adire* cloth with hand-drawn design
190 × 173 cm (75 × 68 in)

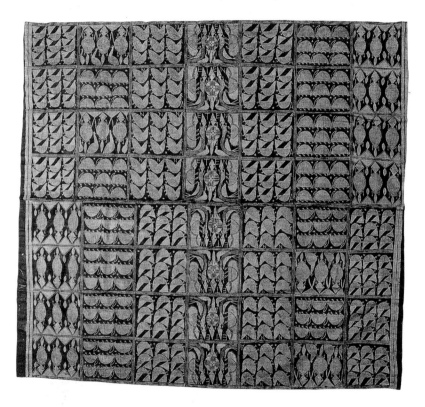

RATHER THAN BEING PURELY DECORATIVE, THE DESIGN MOTIFS ON AFRICAN
TEXTILES ARE OFTEN INTENDED TO PLAY A PROTECTIVE OR SYMBOLIC ROLE.
THE STYLIZED LIZARDS, BIRDS AND RODENTS ON THIS CLOTH HAVE MYTHIC
OR MAGICAL SIGNIFICANCE FOR THE YORUBA.

RESIST-SEWN CLOTH
of the Leopard Society of Cross River
242 × 177 cm (95 × 69½ in)

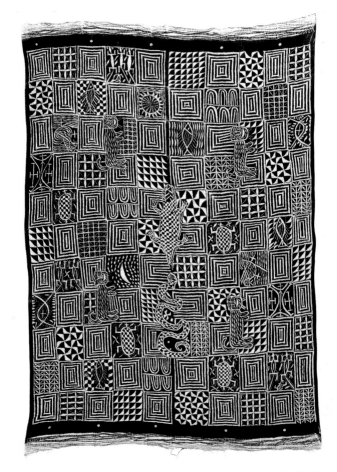

CLUTTERED WITH GEOMETRIC AND ANIMAL SHAPES, THIS
VIBRANT CHEQUERBOARD OF A DESIGN IS RICH WITH
SYMBOLISM. MANY OF ITS MOTIFS ARE THOUGHT TO
POSSESS SPECIAL POWERS.

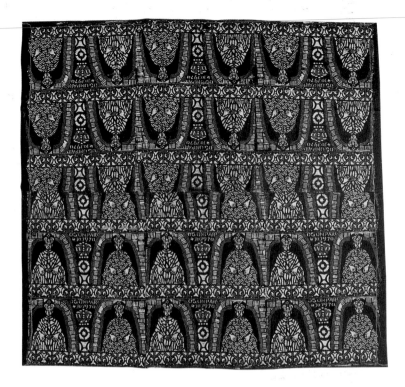

ADIRE CLOTH
indigo-dyed with stencilled design
162 × 179 cm (63¾ × 70½ in)

A FORMAL GRID CONTAINING IMAGES OF A RULER OF ABEOKUTA, DRESSED IN DIFFERENT STYLES OF ROBE. THE ROYAL THEME RECALLS AN *ADIRE* DESIGN POPULAR IN THE FIRST HALF OF THE TWENTIETH CENTURY. THIS WAS INSPIRED BY A BRITISH EMPIRE-WIDE ISSUE OF POSTAGE STAMPS AND MEMORABILIA CELEBRATING THE SILVER JUBILEE OF GEORGE V.

WOMAN'S WRAP
Indigo-dyed *adire* cloth with stencilled design.
Yorubaland, Nigeria.
173 × 194 cm (68 × 76 in)

THIS REPEAT PATTERN OF TOY-
LIKE ELEPHANTS IS UNUSUAL IN
ITS USE OF AN INSCRIPTION
BENEATH EACH MOTIF.

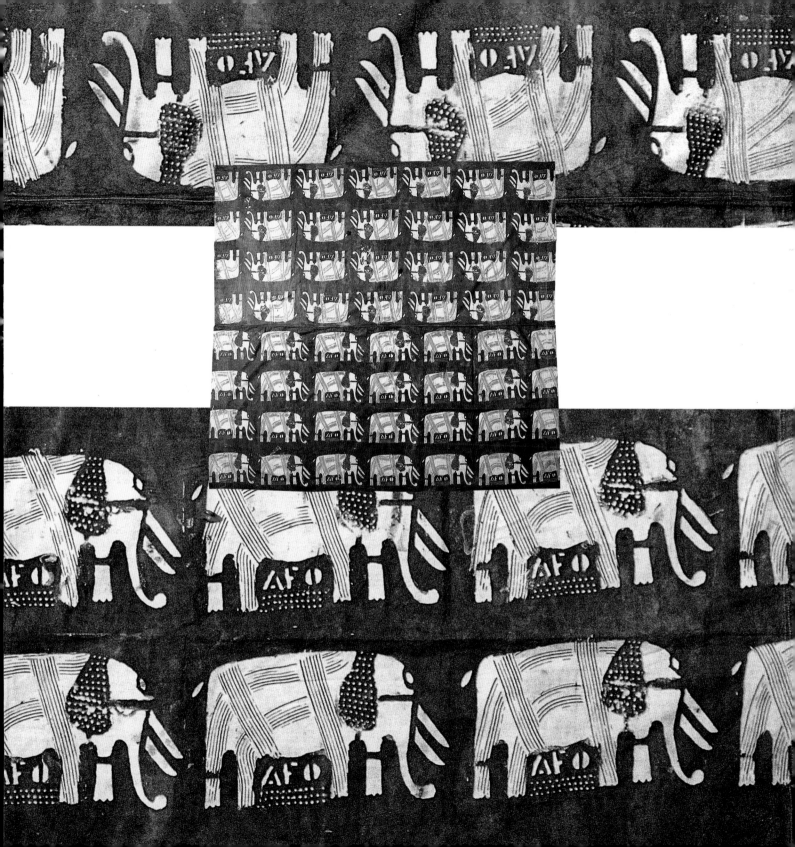

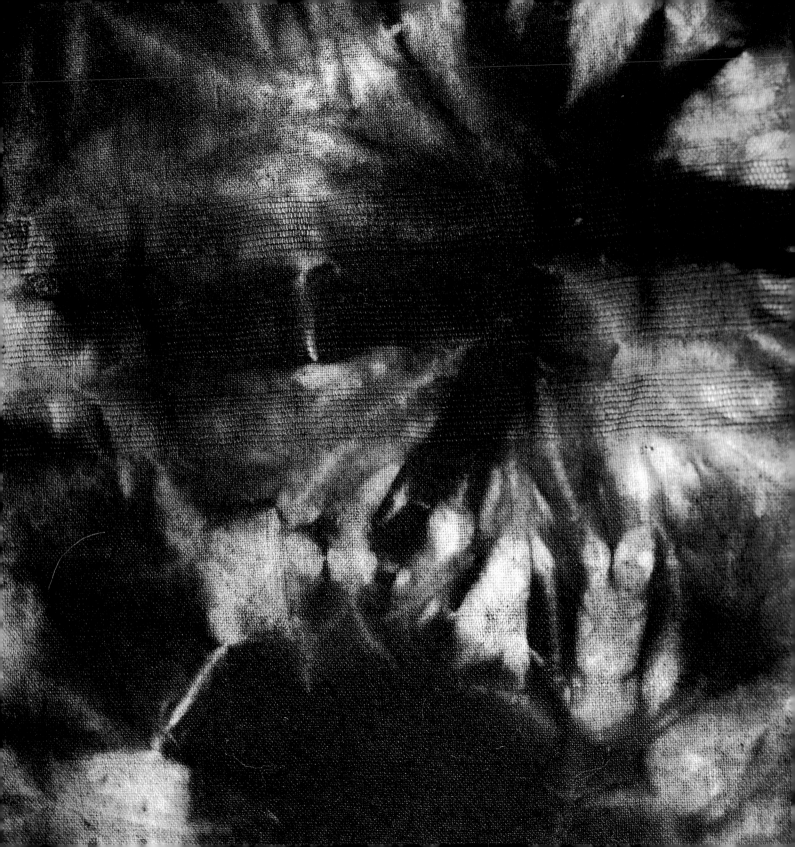

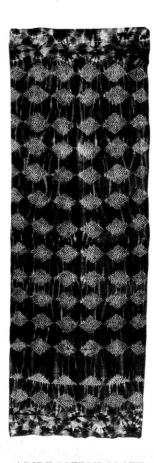

ADIRE COTTON CLOTH
indigo-dyed with both hand-sewn and tied resists
192 × 68 cm (75½ × 26¾ in)

IN SHARP CONTRAST WITH THE FORMAL STITCHED-RESIST LOZENGES THAT MAKE UP THE CENTRAL

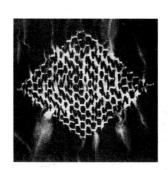

DESIGN, AN EXPLOSION OF RANDOMLY TIED MOTIFS IS USED TO DECORATE EACH END OF THE CLOTH.

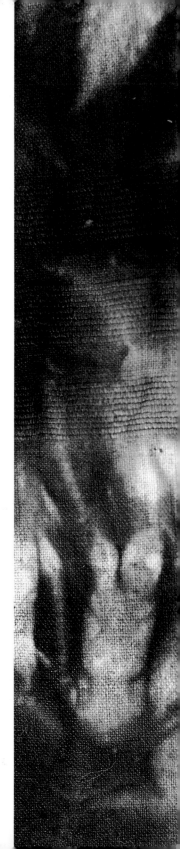

glossary

adinkra traditional fabric of Ghana, decorated with symbolic designs printed on by hand with stamps carved out of CALABASH gourds

adire Yoruba word for resist dyeing with indigo

   adire alabere Yoruba stitched-resist dyeing technique

   adire eleko Yoruba technique whereby designs are painted or stencilled on to the cloth in a RESIST substance (starch paste) before dyeing it with indigo

   adire oniko Yoruba dyeing technique using tied resists, usually of raphia

bakhnug one of three types of decorated shawl worn by women in the mountain areas south of Gabes, Tunisia

barkcloth traditional cloth made by beating the inner bark from particular trees (in Africa, tropical fig trees)

batik Javanese method of resist-dyeing using wax

Berber member of a large ethnic group who have inhabited most of North Africa for more than two millennia

block printing printing a pattern on to cloth by means of wooden or metal blocks into which a design has been carved or engraved

bogolanfini mud-dyed ritual cloths of the Bamana people and other ethnic groups in Mali

calabash dried hollowed-out shell of a gourd from tropical trees of the same name; used for various purposes, including carrying water and other liquids

calligraphy cloth Islamic-influenced cloth decorated with symbols based (often loosely) on Arabic calligraphy

cassava starch derived from the tuberous root of tropical plants of the genus *Manihot*, which are also called cassava (or manioc)

clamp resist dyeing method whereby a RESIST is created by applying pressure to the cloth to be dyed with some kind of vice or clamp

cola nut nut from the *Cola nitida* tree which, when crushed, yields a golden-brown dye; also chewed as a mild narcotic

cut-pile style of embroidery common among the Kuba people of the Congo. Similar to making candlewick, among the Kuba the technique consists of taking raphia threads through the surface of raphia fabric and trimming them off close to the surface with a sharp blade, which produces a velvety pile. Kuba cut-pile cloth is often referred to as Kasai velvet.

damask self-patterned satin-weave fabric in which the warps float over the wefts before being bound in on the pattern areas, and under the wefts of the background, contrasting the two faces of satin; the fabric takes its name from the city of Damascus, where it first came to Western notice

faggoting decorative embroidery technique for stitching two pieces of fabric together selvedge to selvedge

gentian violet a synthetic violet-coloured dye, also used medically for its antiseptic properties

heddle essential part of a loom, used to create the shed openings through which the weft threads are passed

ikat resist process whereby, prior to dyeing, designs are reserved on warp or weft yarns by tying off small bundles of threads with fibre

kente ceremonial strip-woven cloth of the Ashanti people of Ghana

ketfiya general term for the small rectangular shawls worn in the area of Gabes in Tunisia

Kuba a confederation of peoples centred around the Kasai river area in the Congo

lafun cassava or corn starch used by the Yoruba as a resist medium in the making of hand-drawn or stencilled ADIRE cloths *lamba* untailored shawl common in Madagascar

Maghreb (meaning 'the West' in Arabic) North Africa, usually taken as excluding Egypt

mordant a metallic salt that reacts chemically with a dyestuff to fix the dye so that it is permanent

resist a substance (such as starch or wax) or technique (such as tying or stitching) employed to prevent dye from penetrating certain areas of a fabric, in order to create a design in contrasting colours

strip-woven made up by sewing narrow strips of woven cloth together, selvedge to selvedge; strip-woven fabrics are found throughout West Africa

substantive dye one that does not require a MORDANT to render it permanent

supplementary weft technique of applying decorative effects to a fabric by inserting additional (supplementary) weft threads during weaving

tajira multicoloured tie-dye shawl worn as a hair-covering by Berber women to the south of Gabes in Tunisia

tie-and-dye (or tie-dye) a widely used RESIST technique that involves enclosing portions of a fabric within tightly drawn thread ties to prevent them from taking up the dye; it results in (usually circular) patterns that stand out in the original colour of the cloth against a dyed background

tritik resist process whereby designs are reserved by gathering and sewing the cloth before dyeing

ukura term for pictorial indigo-dyed stitch-resist cloths made in northern Igboland for such organizations as the Leopard Society of the Cross River area of south-east Nigeria

warp-faced type of weave in which the warp threads are so closely packed together that they completely obscure the weft threads

Yoruba large ethnic group living in south-west Nigeria

selected reading

Adams, M. and T. R. Holdcraft, 'Dida Woven Raffia Cloth from Côte d'Ivoire', *African Arts*, vol. 25, no. 3, 1992.

Adler, P. and N. Barnard, *Asafo! African Flags of the Fante*, 1992.
African Majesty: the Textile Art of the Ashanti and Ewe, 1992.

Arnoldi, M. and C. M. Kreamer, *Crowning Achievements: African Arts of Dressing the Head*, 1995.

Balfour-Paul, J., *Indigo*, British Museum Press, 1998.
'Muddy River Blues', *Hali*, issue 105, London, 1999.

Besancenot, J., *Costumes of Morocco*, 1988.

Carey, M., *Beads and Beadwork of East and South Africa*, 1986.

Clarke, D., *The Art of African Textiles*, Grange Books, 1997.
African Art, 1997.

Cole, H. and C. Aniakor, *Igbo Arts: Community and Cosmos*, 1984.

Coquet, M., *Textiles africains*, 1998.

Courtney-Clarke, M., *African Canvas*, 1990.

Drewal, H. and J. Mason, *Beads, Body and Soul: Art and Light in the Yoruba Universe*, 1998.

Drewal, H. and J. Pemberton, *Yoruba: Nine Centuries of African Art and Thought*, 1998.

Eicher, J., *Nigerian Handcrafted Textiles*, 1976.

Fagg, W., *Yoruba Beadwork*, 1980.

Fondation Dapper, *Au Royaume du signe — appliqués sur toile des Kuba, Zaïre*, 1988.

Gillow, J. and B. Sentance, *World Textiles: A Visual Guide to Traditional Techniques*, Thames & Hudson, London, 1999.

Harris, J. (ed.), *5000 Years of Textiles*, 1993.

Heathcote, D., *The Arts of the Hausa*, 1976.

Himmelheber, H., *Zaire 1938/39*, 1993.

Idiens, D. and K. G. Ponting (eds), *Textiles of Africa*, Bath, 1980.

Jereb, J., *Arts and Crafts of Morocco*, 1995.

Lamb, V., *West African Weaving*, Duckworth, London, 1975.

Lamb, V. and J. Holmes, *Nigerian Weaving*, Roxford, Hertingfordbury, 1980.

Lamb, V. and A. Lamb, *Au Cameroun — Weaving/Tissage*, Roxford, Hertingfordbury, 1981.

Mack, J., *Malagasy Textiles*, Shire, Aylesbury, 1989.
Emil Torday and the Art of the Congo, 1900–1909.

Menzel, B., *Textilien aus Westafrika*, Museum für V. Ikerkunde, Berlin, 1972.

Murray, J. (ed.), *Cultural Atlas of Africa*, 1981.

Phillips, T. (ed.), *Africa: The Art of a Continent*, 1999.

Picton, J., *The Art of African Textiles*, Barbican Art Gallery, 1995.

Picton, J. and J. Mack, *African Textiles*, British Museum Press, 1995.

Reswick, I., *Traditional Textiles of Tunisia*, 1985.

Sandberg, G., *Indigo Textiles: Techniques and History*, 1989.
The Red Dyes: Cochineal, Madder and Murex Purple, 1994.

Schaedler, K. F., *Weaving in Africa, South of the Sahara*, Panterra-Verlag, Munich, 1987.

Sieber, R., *African Textiles and Decorative Arts*, MOMA, New York, 1972.

Spring, C., *African Textiles*, 1989.
African Arms and Armour, 1993.

Spring, C. and J. Hudson, *North African Textiles*, British Musem Press, 1995.

Stone, C., *The Embroideries of North Africa*, 1985.

museum accession numbers

PAGE	ACC. NO.
2	2000 AF3.1
4	1956 AF27.10
7	1998 AF 1.132 (TOP)
	1973 AF 28.7 (LEFT)
	1934 AF 3-7.132 (RIGHT)
	1964 AF 2.45 (BELOW)
22	2000 Af 3.1
26	1928.61
29	1954+23
30	+5728
32	1963 Af 13.37
35	1930 5-7.16
36	1987 Af 7.8
39	1984 Af 7.14
40	1956 Af 27.10
42	1969 Af 37.2
45	1936 12-11.5
46	1951 Af 3.1
51	1998 Af 1.132
52	1973 Af 28.7
57	1973 Af 28.2
58	1934 3-7.238
61	1949 Af 46.240
63	1953 Af 17.7
64	1964 Af 2.45
66	1971 Af 35.17
71	1934 3-7.132
73	1971 Af 35.19
74	1983 Af 34.1
76	1971 Af 35.26
78	1964 Af 2.45
81	1934 3-7.288
INSIDE COVER	1934 Af 3-7.286

publisher's acknowledgements

The textiles featured in this book are drawn from the collections of the British Museum's Department of Ethnography and have been selected from the viewpoint of their design and technical merit.

We should like to express our thanks to the many people who have helped us in the production of this book, and in particular from the Museum staff: Helen Wolfe, Anna Gaudion and Mike Row. Paul Welti, the art director, must be credited not only for his arresting juxtaposition of illustrations and text, but also for his contribution to the captions analyzing the designs.

picture credits

All pictures are the copyright of The Trustees of the British Museum with the exception of the following: Pages 9, 13, 14, 15, 17 and 19 John Gillow.

author's acknowedgements

The author would like to thank Robert Clyne who so generously introduced me to Nigeria, sharing with me his deep knowledge of Nigerian textiles and correcting many of my misconceptions. I also would like to thank Jenny Balfour-Paul, Brian Durrans, Caroline Hart, Harvey, Ann Hecht, Sheila Paine, Caroline Stone, Kate Wells and Helen Wolfe.

PREVIOUS PAGE: Tie-dyed raphia from the Congo. (See pages 28-9)

index